Traffic Anomaly Detection

Series Editor
Abdelhamid Mellouk

Traffic Anomaly Detection

Antonio Cuadra-Sánchez
Javier Aracil

First published 2015 in Great Britain and the United States by ISTE Press Ltd and Elsevier Ltd

ISTE Press Ltd
27-37 St George's Road
London SW19 4EU
UK

www.iste.co.uk

Elsevier Ltd
The Boulevard, Langford Lane
Kidlington, Oxford, OX5 1GB
UK

www.elsevier.com

British Library Cataloguing-in-Publication Data
A CIP record for this book is available from the British Library
Library of Congress Cataloging in Publication Data
A catalog record for this book is available from the Library of Congress
ISBN 978-1-78548-012-6

Printed and bound in the UK and US

Contents

Introduction

In this book, we show an overview of traffic anomaly detection analysis, which allows us to monitor the security aspects of multimedia services. This approach is based on the analysis of time aggregation adjacent periods of the traffic. As traffic varies throughout the day, it is essential to consider the concrete traffic period in which the anomaly occurs. In this book, we present the algorithms proposed for this analysis. In addition, we make an empirical comparative analysis of these methods and produce a new information theory-based technique which we call "typical day analysis".

In Chapter 1, we present the change point detection algorithms that we are considering in this book. We introduce the CUmulative SUM (CUSUM) control chart, two tests of goodness-of-fit (Pearson's Chi-squared and Kolmogorov–Smirnov tests) and mutual information (mutual dependency between variables). These algorithms are used in the subsequent chapters to detect traffic anomalies in multimedia traffic.

In Chapter 2, we study the periodic behavior of traffic in order to determine significant traffic variations that may reveal how changes in time are statistically significant, in order to find the optimal aggregation period. The chapter settles the basis for determining the optimum period of multimedia traffic aggregation, for network monitoring purposes. The proposed approach considered the optimal interval to be the one which maximizes the proportion of periods showing substantial statistical changes, and for this we used Pearson's Chi-square and Kolmogorov–Smirnov tests of goodness-of-fit.

In addition, in Chapter 3, we analyze how the different algorithms behave in detecting changing points. We present a deep analysis of the multimedia monitored traffic to understand how the traffic behaves throughout the day to compare the traffic anomaly detection methods. This approach is more useful than traditional sudden peak change techniques as we consider the concrete traffic period in which the anomaly occurs, as traffic behaves differently throughout the day.

Finally, in Chapter 4, a new information-theory technique is proposed: we present the "typical day profile" technique and its application to change point detection field. This constitutes a new information-theory-based technique that analyzes the traffic pattern over a typical 24 h day. After carrying out a deep traffic analysis to be aware of normal traffic behavior, we adjust our procedure by considering which combination of algorithms gets better performance in each period of the day. The analysis we have performed confirms that a combination of the algorithms provides better results than the use of a single one. Therefore, we conclude that there is no one single method that is highlighted, but a combination of different algorithms depending on the underlying traffic and the time of the day provides much more accuracy in detecting traffic changes.

1

Introduction to Traffic Anomaly Detection Methods

In this chapter, we describe the different change point detection algorithms that we are considering in this book. We present the CUmulative SUM (CUSUM) control chart, two tests of goodness-of-fit (Pearson's Chi-squared and Kolmogorov–Smirnov (K-S) tests) and mutual information (MI) (mutual dependency between variables). Such algorithms will be used in the following chapters to detect traffic anomalies for different purposes.

1.1. Cumulative sum control charts (CUSUM)

Statistical control charts (SCCs) perform measurements obtained as variations from the expected value by the standard deviation. In particular, CUSUM (CUMulative SUM) is a sequential analysis algorithm that allows us to monitor sudden changes in continuous processes, such as changes in traffic.

The state of the art shows that there is no single SCC algorithm that stands out from the other ones, but the performance of SCC depends on the underlying traffic characteristics [MAR 11, MAT 11, CAR 12, OPR 13, BUL 12].

So, in order to simplify this study, we have focused on the most renowned one, the CUSUM chart, as the implementation of other control chart methods (e.g. Exponentially Weighted Moving Average (EWMA)) would lead to equivalent results.

CUSUM consists of a cumulative sum whose value C_i determines if the process is under control. When CUSUM exceeds a certain threshold value, it is then considered that a change has occurred.

The CUSUM chart plots the cumulative sums of the deviations of the sample values from a target value [MON 04]. If μ_0 is the target for the process mean, then the cumulative sum control chart is:

$$C_i = \sum_{j=1}^{i} (\widetilde{x}_j - \mu_0) \qquad [1.1]$$

In order to determine if the CUSUM values are under control, two thresholds C+ and C (called one-sided upper and lower CUSUMs, respectively) are defined:

$$C_i^+ = max[0, x_i - (\mu_0 + K) + C_{i-1}^+] \qquad [1.2]$$

$$C_i^- = max[0, (\mu_0 - K) - x_i + C_{i-1}^-] \qquad [1.3]$$

where the starting values are $C_0^+ = C_0^- = 0$ and K is a reference value that is typically chosen about halfway between the target μ_0 and the out of control value of the mean μ_1 that we are interested in detecting quickly. If the shift is expressed in standard deviation units as $\mu_1 = \mu_0 + \delta\sigma$, then K is one-half the magnitude of the shift or

$$K = \frac{\delta}{2}\sigma = \frac{|\mu_1 - \mu_0|}{2} \qquad [1.4]$$

Figure 1.1 represents the values of C_i^+ and the threshold. When the values of C_i^+ surpass the threshold, the process is out of control (last two samples in the example).

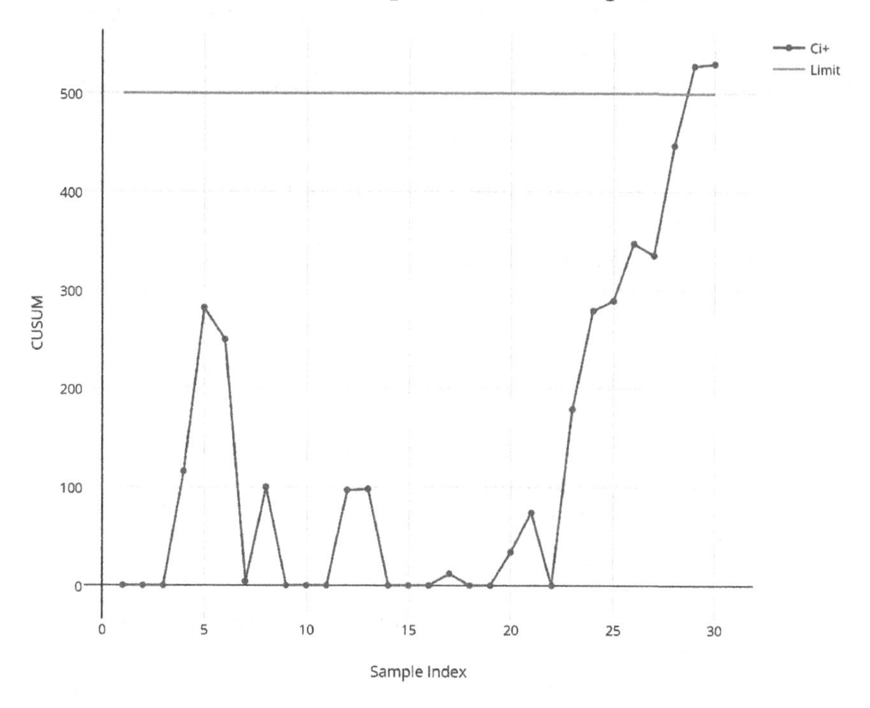

Figure 1.1. *Example of CUSUM control chart*

The SCCs are graphs that show whether a sample of data falls within the normal range of variation. In CUSUM (also known as CSUM) as measurements are taken, the difference between each measurement and the benchmark value is calculated, and this is cumulatively summed up. When there is an anomaly in the traffic, the CUSUM value will progressively depart from that of the benchmark.

For our purposes, all the values from a period are compared to those from the immediate previous period.

1.2. Tests of goodness-of-fit

The tests of goodness-of-fit [CRO 12] are used to derive whether a given time interval of 1 day shows a traffic distribution independent from the rest of the intervals (i.e. how similar two consecutive periods are) that can be used to determine changes.

In this work, the two main tests of goodness-of-fit have been carried out. Such tests are described in the following sections: Pearson's Chi-square test [CHE 54] and Kolmogorov–Smirnov (K-S) test [PET 77].

1.2.1. *Pearson's Chi-squared test (χ^2)*

This test measures the discrepancies between the expected number of times each outcome occurs (assuming that the model is true) and the observed number of times each outcome occurs [PEA 00].

The goodness-of-fit is determined by comparing the observed values with those expected [PEA 00]. This study considers that the observed and expected values of a sample are related to a time interval and the previous one. Specifically, it compares the values of the previous hour with the current one in this way.

$$\chi_j^2 = \sum_{i=1}^{k} \frac{(H_i^{j-1} - H_i^j)^2}{H_i^j} \qquad [1.5]$$

where:

– j is the time interval (i.e. hour);

– k is the number of days;

– H_i^j is the aggregate value of j interval;

– H_i^{j-1} is the aggregate value of the previous interval (j-1).

The decision rule to determine the critical value in this case considers the hypothesis that H_0 is rejected if the calculated statistical value is greater than or equal to the theoretical value found from the table of the Chi-square distribution:

– if $\chi^2_{calc} \geq \chi^2_{\alpha}$, then reject the null hypothesis;

– if $\chi^2_{calc} < \chi^2_{\alpha}$, then accept the null hypothesis;

where χ^2_{calc} is the statistic value obtained and χ^2_{α} is the threshold for a given α as shown in Figure 1.2.

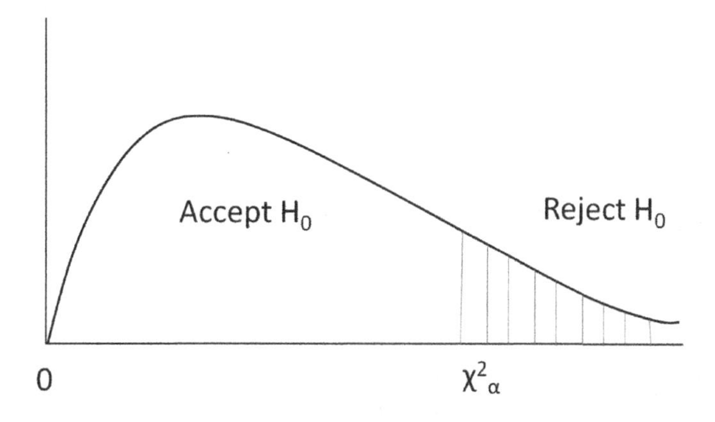

Figure 1.2. *Accepted and rejected regions for χ^2 test*

First, the statistical values from added traffic are calculated for each interval. The Chi-squared test measures the discrepancies between the expected number of times each outcome occurs (assuming that the model is true) and the observed number of times each outcome occurs.

In this work, it is considered that the observed values correspond to the current aggregate time period, and expected the preceding period.

To calculate the statistics of each interval, we create a histogram with the number of occurrences for different days.

To ensure the goodness of the results, each cube (segment of the histogram) should contain at least six samples [HIG 04].

In order to determine whether to reject or accept the hypothesis, we use the critical values. The critical threshold is determined depending on the degrees of freedom (number of cubes minus one [HIG 04]) and alpha.

For example, for $\alpha = 0.05$, the tabulated value of the Chi-squared distribution with 7 degrees of freedom is equal to 14.067 [HIG 04]. Thus, the hypothesis is rejected if $\chi^2 < 14.067$.

These intervals have a similar probability distribution within these periods and independent of adjacent periods, well differentiated from the rest. This allows us to detect changes in network behavior in these time slots.

1.2.2. *Kolmogorov–Smirnov test*

This test [PET 77] checks that the distribution of a set of samples conforms to the theoretical distribution.

It measures the maximum distance D between two consecutive cumulative distribution functions (CDFs) [PET 77], as shown in Figure 1.3.

In our environment, this test compares the CDFs of each period with the previous one.

$$D_j = max|F_{j-1}(x) - F_j(x)| \quad [1.6]$$

where:

– j is the time interval (i.e. hour);

– $F_j(x)$ is the CDF of j interval;

– $F_{j-1}(x)$ is the CDF of the previous interval (j-1).

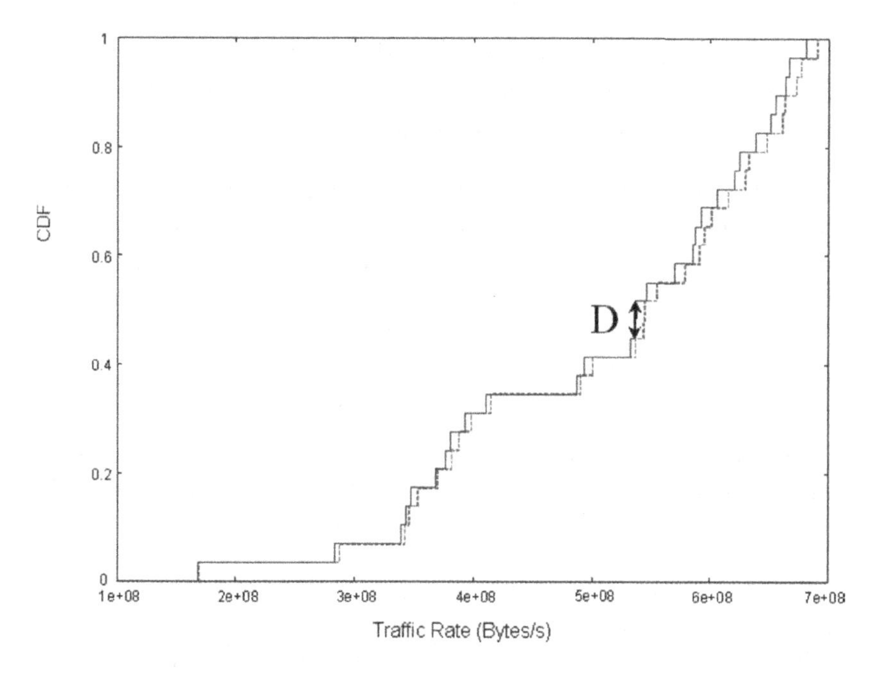

Figure 1.3. *D value to determine the K-S test of goodness-of-fit. For color version of the figure, see www.iste.co.uk/cuadra/traffic.zip*

First, we determine the threshold above which it is considered that the statistical value does not meet expectations, that is the null hypothesis is rejected for α if:

$$\sqrt{\frac{nn'}{n+n'}}D_{n,n'} > K_\alpha \qquad [1.7]$$

where D is the K-S statistic value and n, n' are the samples ($n = n' =$ number of days) [PET 77].

The K_α values are tabulated in [HIG 04]. For $\alpha = 0.05(95\%)$ and $n = 54$, $K_\alpha = 0.18$.

To test the hypothesis, the two CDFs are compared, that is the empirical and theoretical distribution function.

When comparing empirical values, the observed and expected CDFs are used; in this work the CDF of one period and the immediately preceding one.

Once both distributions are calculated, the K-S statistic is determined from the biggest difference between both functions, the maximum distance between two consecutive CDFs.

1.3. Mutual information (MI)

The MI is a measure of the amount of information that one random variable contains about another one. This means that if both random variables are independent, the MI is zero.

The MI [BEL 62] of two random variables is a quantity that determines the mutual dependence of the two random variables [SHA 01] and measures the reduction in uncertainty of a random variable, X, due to another variable, Y [MAR 12].

The MI of two random variables, X and Y, with the joint probability distribution function $p_{X,Y}(x, y)$, is defined as:

$$I(x, y) = \sum_{x}\sum_{y} p_{X,Y}(x, y) log \frac{p_{X,Y}(x, y)}{p_X(x)p_Y(y)} \qquad [1.8]$$

The former equation can also be expressed in terms of entropy. The information that a period has about the other, or the MI, can be defined by the following equation:

$$I(X; Y) = H(X) + H(Y)H(X, Y) \qquad [1.9]$$

The information that period X has about Y (or Y has about X as information is a symmetrical quantity) is the difference between the sum of the entropies of each and the joint entropy (H (X,Y)) as shown in Figure 1.4.

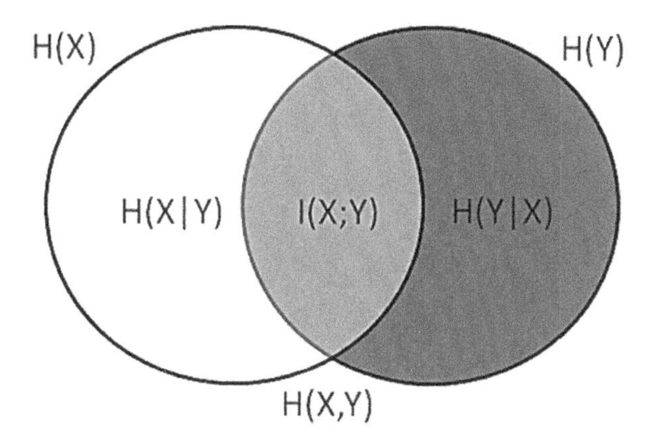

Figure 1.4. *Representation of mutual information based on entropy*

In this work, MI is used as a measure of the similarity between the traffic of two consecutive intervals. This way, the MI measures the amount of information that one period of traffic contains about the previous one and therefore constitutes an algorithm to detect changes in traffic.

Since there are no reference values in terms of figures that can take the MI algorithm, beyond the minimum is zero when they are independent variables, typically a value of normalized MI (MIn) with respect to the highest value that is taken in a series of traffic is used.

2

Finding the Optimal Aggregation Period

In recent years, several studies that analyze the behavior of Internet traffic have been developed, driven by the growing demand for online applications during the last decade. The longitudinal analysis of traffic helps us to examine how the network behaves from a temporal evolution, in particular, the traffic carried on the Internet. The state of the art shows that the choice of an aggregation period for Internet traffic is normally just based on intuition, well-known concepts inherited from telephone networks such as the busy hour. However, there is no justification on whether the aggregation period chosen is optimal or not. By optimal, we mean that the aggregation period is large enough to reduce the data set (for example to the order of minutes) but with minimal impact in the information loss due to aggregation from the original time series. The purpose of this chapter is to study the periodic behavior of traffic in order to determine significant traffic variations that may reveal how changes in time are statistically significant. We propose to use tests of goodness-of-fit to identify significant variations in the distribution of traffic among time intervals. In this chapter, we determine the optimal aggregation time interval based on traffic characteristics for highly multiplexed Internet traffic collected from a campus university which differs from traditional busy-hour periods. Traffic profiles of a "typical day" are highly relevant to detect deviations from "normal" behavior, whose disorders can help us to detect traffic anomalies in the network.

2.1. Introduction

Traffic studies are an essential source of information to understand not only the current status of the Internet traffic, but its evolution as well. The main works on traffic analysis are focused on obtaining measurements of Internet traffic in the medium term [MCC 00, FOM 04, MAD 06] (a longitudinal perspective of traffic with continuous traces of several weeks) in order to determine the behavior of the Internet [CAL 10] and its users [GAR 11], which will enable development of more reliable trend analyses. These studies are extremely rich but very complex because of the need for a constant feed of traffic from weeks to several years, the difficulties of accessing real user traffic, and the necessity of data aggregation for both computational purposes and exploitation of results for traffic monitoring tasks. The choice of an aggregation period for Internet traffic is normally based on intuition, and it is based on well-known concepts from telephone networks such as the busy hour. However, there is no justification on whether the aggregation period chosen is optimal or not. By optimal, we mean that the aggregation period is large enough to reduce the data set (for example to the order of minutes) but with minimal impact in the information loss due to aggregation from the original time series.

In order to benchmark the different aggregation periods, we choose to identify aggregation periods that provide "different" marginal distributions, i.e. distributions with significant distance between each other. For example, for an aggregation period of one day, the underlying marginal distributions would be very similar and the information provided would also be very basic, with variations between weekdays and weekends. However, as the aggregation period gets reduced, the marginal distributions offer a better portrait of the Internet traffic time series. Eventually, for a very small timescale (say, milliseconds) the marginal distribution provides very detailed information, but the aggregation is too poor. We wish to find a trade-off between information and data reduction in the time series. To this end, we proceed with a typical aggregation period of 60 min, and reduce the aggregation period until the underlying distributions are different in the probabilistic sense and thus provide information which is not contained in a coarser aggregation period. To measure the degree of similarity of two distributions, we use goodness-of-fit statistics such as the well-known Pearson's χ^2 test and the Kolmogorov–Smirnov statistic.

In this chapter, we determine the optimal period of aggregation from traffic characteristics. The optimal period in network management field is the one that allows us to monitor the network behavior per interval, without losing any essential information; that is, to monitor the network using that interval to detect as many changing situations as possible, which ultimately constitute evidence of network anomalies. As shown in the state of the art, an arbitrary criterion has been used for temporal aggregations so far, traditionally using the *de facto* period for analog telephony networks (busy hour).

Our proposal and the main objective of this work is to use the significant variations in the distribution of traffic between intervals as a decision criterion when determining the best interval period. Therefore, we propose to maximize the proportion of periods that present relevant changes between intervals at statistical level, which are obviously a symptom of changing situations (which, as we have introduced, determine the optimal period of aggregation) quantified by using tests of goodness-of-fit. We have established a procedure to determine the optimal period of aggregation that has been applied to the time series of traffic of 58 consecutive days from a university campus in Spain that gathers both uplink and downlink Internet traffic collected from RedIRIS [GAR 11] (the Spanish National Research and Education Network, NREN). As we present in this book, we conclude that the optimal traffic aggregation periods for that traffic series are 10 min downlink and 15 min uplink.

The following sections of this chapter are structured in this way, starting with a deep study of the state of the art and issues not sufficiently covered by the current literature. The third section summarizes the macroscopic observation of the highly multiplexed Internet traffic where the independence of the traffic is proved. The chapter continues with the average-day analysis section where we show the optimal period for all intervals. Finally, the last section includes the conclusion of this work.

2.2. State of the art

In recent years, there have been some studies that analyze the behavior of Internet traffic [MAI 09, PAX 96, WIL 98, BRO 02]. This section analyzes the state of the art at the microscopic level (short-term "typical day"), which examines the traffic throughout one day and its alterations in shorter periods (hours or less). Moreover, we describe those aspects not covered by the current situation that we have developed in this work. In [GAR 11], the authors perform a medium-term traffic comparative between different days using the busy hour as the chosen period. The traffic capture took place in the RedIRIS (the academic and research network Spanish) over four months (January–April 2009). They analyzed both directions of traffic, the uplink (user to network) and downlink (network to user). The article shows that the observed traffic during peak hours may be modeled by a white Gaussian process. This implies that the network traffic at these times can be parametrized by the mean and standard deviation. The main conclusions are that the network size (in terms of number of users) is the main cause of traffic load in the busy hour. The authors established a linear regression model that adjusts the amount of traffic for each user on the values of the busy hour. Other work that reflects traffic studies during time periods, in this case 90 min [PAP 05], is summarized here. The authors present a methodology to predict when and where to establish new links in the core of an IP network. First, the aggregate traffic is measured between two neighboring points of presence (PoP) in the core of an IP network, and then the authors analyze its evolution over time scales of 1.5 h (the smallest time scale that allows them to observe the behavior of the links in the time series for the periods of interest). The results show that IP backbone traffic has long-term trends, strong periodicity and variability on multiple time scales. The authors make accurate predictions about the behavior of traffic at small time scales, from one day to another, or between certain hours on different days.

Finally, in [BEI 04], the authors made an analysis of a typical day from aggregate hours in order to study the behavior of users when they view Internet search engines, and later in [BEI 07] they extended the study to days, weeks and months. The paper shows the analysis of hundreds of millions of log records of visits to Web browsers for a consecutive period of six months. The first study [BEI 04] focuses on analyzing the aggregate traffic for different times of day for a week, and shows that there are changes in trends in queries throughout the day, especially in the categories (content type) visited. Later, in [BEI 07], they extended the study to six months and again found that time trends remain stable despite fluctuations in the volume of queries, such as thematic categories in the short term (during business hours) and long term (several weeks or months).

As described in the previous paragraphs, the research based on the analysis of a typical average day is not being sufficiently covered by the state of the art. Thus, to the best of our knowledge, no published work to date raises traffic analysis of a typical day by using tests of goodness-of-fit to prove the independence of the distributions of traffic between the different periods of the day. Those previous studies simply use a standard time of 60 min (90 min in [PAP 05]), traditionally employed as a *de facto* period (from the busy hour voice circuits in analog telephone). However, the authors do not justify the choice of these periods of analysis. In [GAR 11], the authors conducted a study based on the analysis of the busy hour, so traffic directly employs aggregate periods of 60 min, but do not specify the reason for using the busy hour for IP traffic analysis. In [PAP 05], the authors used an arbitrary period of 90 minutes prior to a 12 hour interval analysis. The reason why the authors have chosen this period seems to be to simplify the calculations of later aggregations, as eight periods of 90 minutes cover the aggregate of 12 hours. Finally, in [BEI 04] and [BEI 07], the authors performed studies based on hourly traffic analysis and again do not justify the selection of this period, beyond routine use of this type of aggregate.

This chapter delves into traffic analysis to study the behavior of typical day profiles from aggregated time intervals. To determine the optimal period of analysis, we examine the aggregate traffic from different periods (10 min, 15 min, 30 min, 60 min, etc.) by using two separate tests of goodness-of-fit, the result of which will determine the period that generates the best result in tests. This chapter describes a procedural study that identifies the optimal period for traffic aggregation based on the study of traffic itself. This will comprehensively let us identify moments that present significant changes, such as traffic diversions about "typical days" with normal behavior. This information is useful to know the behavior of the network and detect network anomalies.

2.3. Macroscopic observation of traffic

In this section, we analyze a series of eight-week highly multiplexed traffic flows (58 consecutive days) collected at a campus university. The traffic was monitored in both directions of the access link, i.e. downlink (from the Internet to the user) and uplink (from the user to the Internet). The objective is to determine how traffic behaves during different periods of the day.

On one side, the multiplexed traffic is analyzed through the cumulative distribution function (CDF) of each period for the different days of the sample. In order to determine the variations, we use the most common tests of goodness-of-fit: Pearson's Chi-square test [CRO 10, CHE 54], and Kolmogorov–Smirnov test [PET 77], whose results appear in the next chapter. Both tests of goodness-of-fit require that the multiplexed samples are independent in each period. This way, in this section, we calculate the autocorrelation of traffic. Figure 2.1 summarizes the procedure.

2.3.1. *Traffic overview*

To perform this study, we have arranged a series of Internet traffic flows from a campus university that has around 13,000 students and over 1,200 teachers, that together with the administrative and service staff formed a total of 15,000 Internet users, giving highly multiplexed monitored traffic. This is a total of 57 full days, with traffic uplink and downlink added at intervals of 5 min to have a mean throughput (KBytes/s).

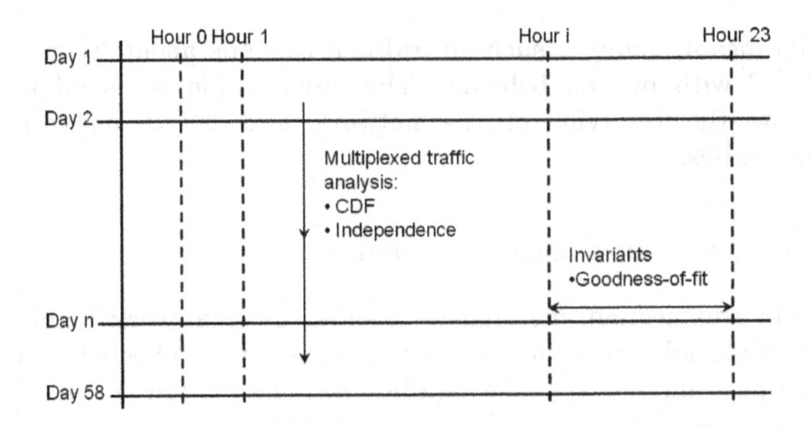

Figure 2.1. *Comparison of traffic between consecutive time periods*

2.3.2. *Observation of traffic*

To determine the characteristics of the multiplexed traffic, a whole series of days for the same period has been analyzed. Therefore, for each hourly aggregation, we have studied a cross-sectional of the days that have contributed to the aggregation. The observation of traffic has been carried out at two levels: through cumulative density functions of each period, and through the autocorrelation of the samples to test the independence of the series, *sine qua non* to confirm the reliability of the goodness-of-fit tests shown in the next section.

2.3.2.1. *Cumulative distribution function*

As a prelude to detailed traffic analysis, we calculate the CDFs [PET 77] for each period. As an example, Figure 2.2 shows a subset of the 24 CDFs, particularly those for periods of 0 h, 6 h, 12 h and 18 h downlink.

2.3.2.2. *Independence test*

This section determines the independence of the traffic samples from the same period (hour h, values $X_1..X_{n_0}$) where n_0 is the last day of the sequence. To this end, the correlogram is calculated from the sequence X_i (traffic in each period, i.e. one hour) by using the method developed by Cox [COX 66] “Estimates of first- and second-order moments of intervals”, where the estimator of the correlation coefficient of the series of interval j is obtained from the covariance estimators also available in [COX 66]:

$$\widetilde{\rho_j} = \frac{\tilde{C}_j}{\left(\tilde{C}'_{0,j}\tilde{C}''_{0,j}\right)^{1/2}} \qquad [2.1]$$

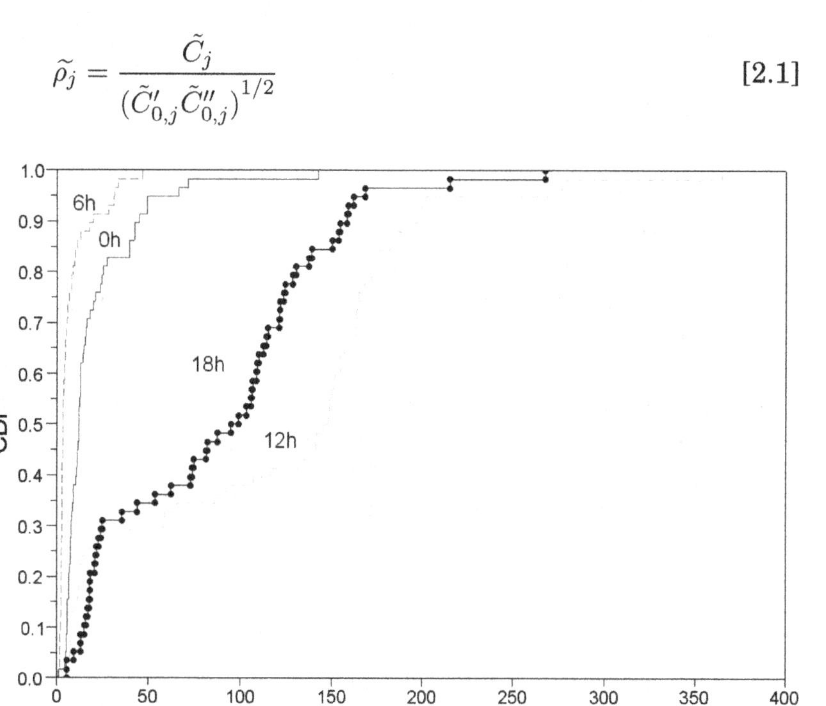

Figure 2.2. *CDFs of 0 h, 6 h, 12 h and 18 h. For a color version of the figure, see www.iste.co.uk/cuadra/traffic.zip*

To determine the degree of independence of the series of the same period, we first calculate the correlation coefficients $\widetilde{\rho_j}$. A preliminary analysis of the correlation coefficients suggests that, due to traffic characteristics, there is a strong correlation every 7 days. Thereby, we do not use those days to ensure a greater reliability of the data, reducing the original data source to 54 full days. To verify the reliability of the estimates of the correlation coefficients, the thresholds on which the values of the correlogram should be delimited are depicted. First, we calculate the asymptotic variance [COX 66] of $\tilde{\rho}_j$ as:

$$\tilde{\Delta} = \frac{1}{2}\left\{1 + 2\sum_{j=1}^{n_0-1}(1 - \frac{j}{n_0})\tilde{\rho_j}^2\right\} \qquad [2.2]$$

To satisfy the independence assumption, the autocorrelation coefficients should be as small as possible, ideally tending to zero. This condition is fulfilled for $\tilde{\rho}_j < C_{1/2}\alpha$ for sufficiently small values of alpha, wherein $C_{1/2}\alpha$ is the percentile $1/2\alpha$.

To determine the value of $C_{1/2}\alpha$, we can use as reference the tabulated values from a normal distribution N (0,1) for a specific accuracy of $\alpha = 0.05(95\%)$ that is translated to the actual distribution $N(\mu, \sigma)$ so that $C_{1/2} = 1.92\tilde{\rho}_j$. As:

$$\tilde{\rho_j} = \frac{\Delta}{(n_0 - j)} \qquad [2.3]$$

We determine the thresholds for each period. We analyzed the aggregated periods of 10, 15, 30, 45, 60, 90 and 120 min in both uplink and downlink. The following figure shows the correlograms of most significant times of day, from 12 h to 15 h, along with their respective thresholds for aggregate hours downlink.

As can be observed, the values of the correlogram are within the threshold for the main lags (five units), but with one minor negative value, which verifies the assumption of independence. This situation is true for the rest of the hours, which are also within the critical region of acceptance. To verify the independence of the samples, we have carried out a study of the estimates of the correlation coefficients for all aggregates (uplink and downlink), for a lag of five units, comparing their respective thresholds. The result is that the first five correlation coefficients are within the acceptance critical region for all intervals analyzed, confirming the hypothesis of independence.

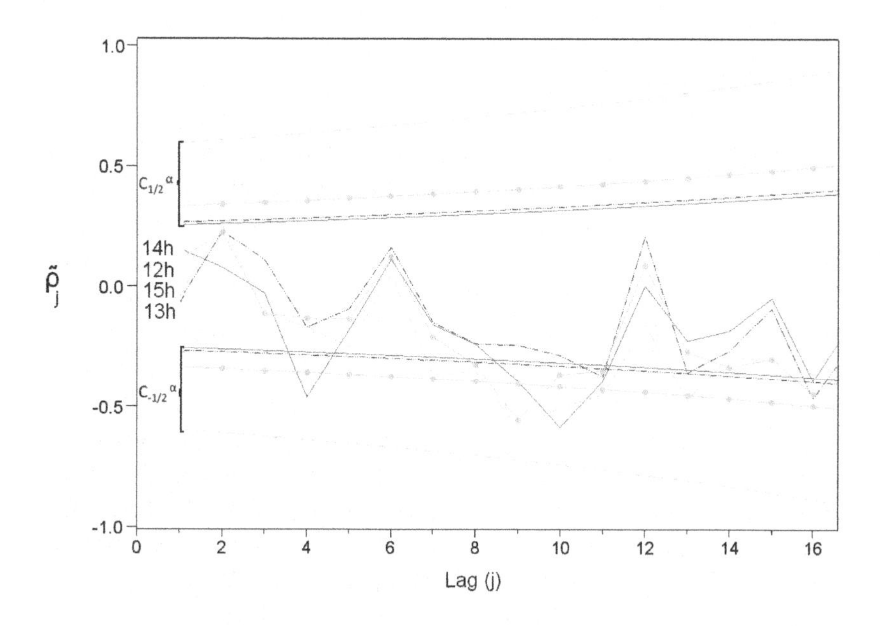

Figure 2.3. *Correlograms of 12 h (red, solid line), 13 h (blue, dotted line), 14 h (green, dash-dotted line), 15 h (cyan, dashed line) and their related thresholds. For a color version of the figure, see www.iste.co.uk/cuadra/traffic.zip*

2.4. Average-day analysis

In this section, we focus on the analysis of a typical day from the aggregated data. Our study is based on the results of the tests of goodness-of-fit that will be used to derive whether a given time interval of one day shows a traffic distribution independent of the rest of the intervals. This helps to identify time slots in which the traffic shows an underlying distribution, which can be used to determine changes.

In determining the goodness of results, two separate tests of goodness-of-fit have been carried out. These tests are described in the following sections: Pearson's Chi-square test [CHE 54] and the Kolmogorov–Smirnov test [PET 77].

2.4.1. *Pearson's Chi-square test*

The goodness of samples is determined by comparing the observed values with those expected [PEA 00]. This study considers that the observed and expected values of a sample are related to a time interval and the previous one. First, the statistical values from added traffic are calculated for each interval. The Chi-square test measures the discrepancies between the expected number of times each outcome occurs (assuming the model is true) and the observed number of times each outcome occurs. In our case, it is considered that the observed values correspond to the current aggregate time period, and expected the preceding period. To calculate the statistics of each interval, we create a histogram with the number of occurrences for different days. To ensure the goodness of the results, each cube (segment of the histogram) should contain at least six samples. In order to determine whether to reject or accept the hypothesis, we use the critical values. The critical threshold is determined depending on the degrees of freedom (number of cubes minus one [HIG 04]) and the accuracy required (α). For example, for $\alpha = 0.05$ (i.e. 95% of accuracy), the tabulated value of the Chi-square distribution with 7 degrees of freedom is equal to 14.067 [HIG 04]. Thus, the hypothesis is accepted if $\chi^2 < 14.067$.

These intervals have a similar probability distribution within these periods and are independent of adjacent periods, well differentiated from the rest. This allows us to detect changes in network behavior in these time slots. The optimal period is determined by the period that maximizes the number of intervals that pass the test of goodness.

The following table shows the number of intervals where the hypotheses are accepted for the aggregate periods of 10, 15, 30, 45, 60, 90 and 120 min (uplink and downlink) and the proportion of intervals that exceed the test of goodness-of-fit (optimal values are in italics).

Period	Intervals	$N\chi^2_{Up}$	$\%\chi^2_{Up}$	$N\chi^2_{Down}$	$\%\chi^2_{Down}$
10	144	86	*59.72*	62	43.056
15	96	47	48.958	45	*46.875*
30	48	20	41.667	21	43.75
45	32	10	31.25	8	25.0
60	24	5	20.833	8	33.333
90	16	3	18.75	2	12.5
120	12	1	8.333	0	0.0

Table 2.1. *Intervals that accept the Chi-square hypothesis*

The aggregate which maximizes the number of intervals for the downlink traffic is 10 min, while uplink is 15 min. Figure 2.4 shows the test of goodness-of-fit of the optimal interval uplink.

2.4.2. *Kolmogorov–Smirnov (K–S) test*

This test measures the maximum distance D between two consecutive CDFs [PET 77].

The critical threshold K_α values depend again on α, and are tabulated in the bibliography. In particular, [HIG 04] includes a table where K_α is represented for different values of alpha. For $\alpha = 0.05(95\%)$ and n = 54, the tabulated statistical critical value $K_\alpha = 0.18$.

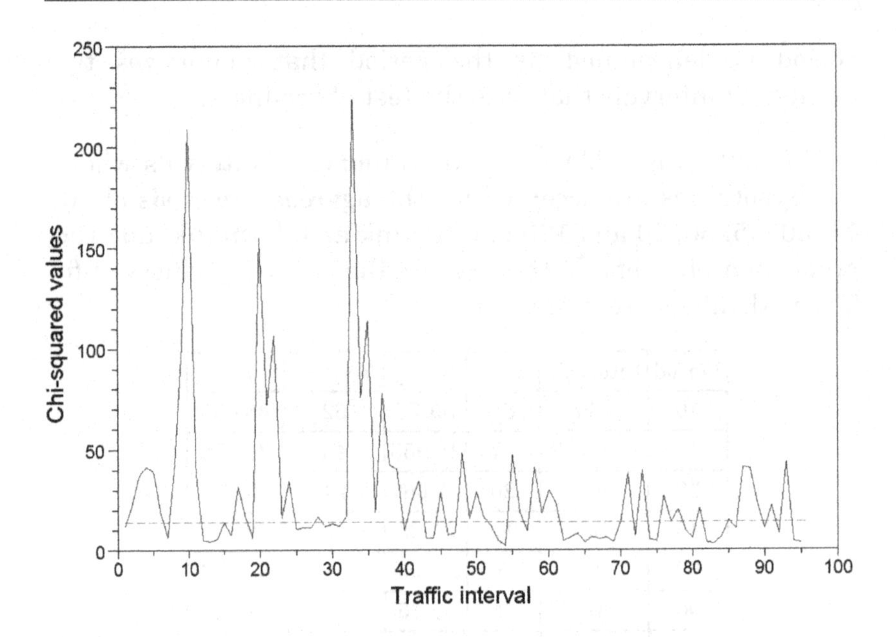

Figure 2.4. *Chi-square statistical values of 15 min aggregate (uplink) and threshold for alpha = 0.05*

The conclusions are equivalent to those obtained from the Chi-square test, namely that these intervals have very similar traffic distribution, clearly differentiated from the rest, identifying variations in behavior in the network. Similarly, the optimal period is the period that maximizes the number of intervals that pass the test of goodness-of-fit.

Table 2.2 shows the number of intervals where the hypotheses are accepted for the aggregate periods of 10, 15, 30, 45, 60, 90 and 120 min (uplink and downlink) and the percentage of intervals that exceed the test of goodness-of-fit. We represent the optimal values in italics.

As shown, the aggregate value which maximizes the number of intervals for the downlink traffic is 10 min, while uplink is 15 min. Figure 2.5 shows the test of goodness-of-fit of the optimal interval downlink.

Period	Intervals	NK_{Up}	$\%K_{Up}$	NK_{Down}	$\%K_{Down}$
10	144	86	*75.694*	114	79.167
15	96	47	65.625	79	*82.291*
30	48	20	54.167	28	58.333
45	32	10	40.625	14	43.75
60	24	5	33.333	9	37.5
90	16	3	18.75	4	25.0
120	12	1	25.0	2	16.667

Table 2.2. *Intervals that accept the K–S hypothesis*

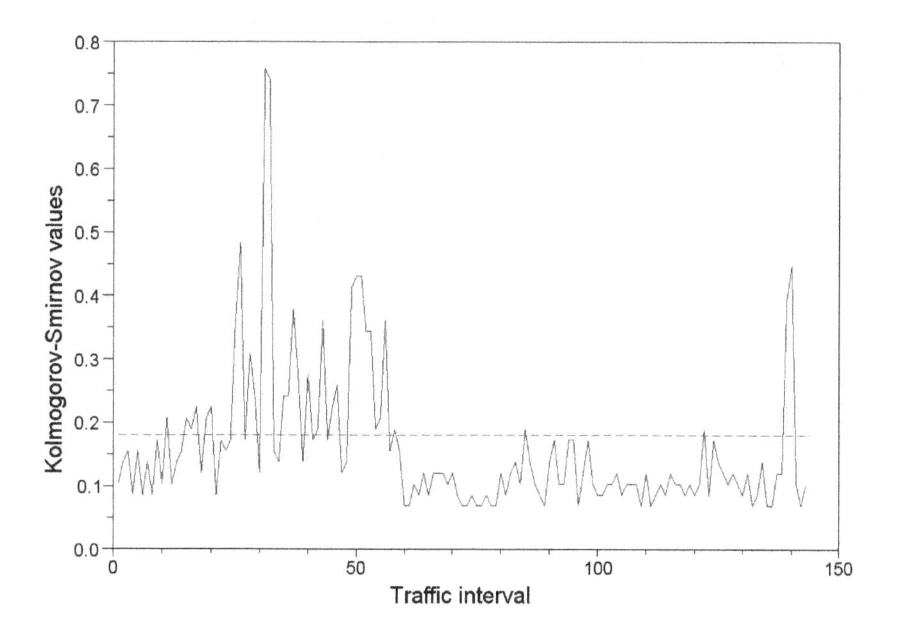

Figure 2.5. *K–S statistical values of 10 min aggregate (downlink) and threshold for alpha = 0.05*

2.4.3. *Detection of the optimal aggregation period*

From the previous test of goodness-of-fit we see that very similar results are obtained in both cases, which proves the accuracy of the results. First, we see that Pearson's

Chi-square test is somewhat more restrictive than the Kolmogorov–Smirnov test, so that there are more intervals where hypotheses are accepted in K–S. In fact the accepted periods are slightly extended from the first test, i.e. K–S results concatenate certain time slots where the hypotheses are accepted.

Beyond the results obtained from the trace of traffic available, we have established a procedure to detect the periods whose traffic distribution are independent of the rest. The purpose of this method is to determine time intervals in which traffic presents a different distribution, with applications such as the detection of changes in network behavior.

Table 2.3 shows the optimal time interval per direction.

Direction	$\%\chi^2$	$\%K-S$	Optimal period (min)
Uplink	46.875	82.291	15
Downlink	59.722	75.694	10

Table 2.3. *Optimal aggregation period*

The optimal intervals match in both tests of goodness-of-fit and we conclude that those are the most optimal aggregated periods for the analyzed traffic.

2.5. Conclusion

In this chapter, we have settled the basis for determining the optimum period of Internet traffic aggregation, for network monitoring purposes, in contrast to the *de facto* aggregation period used in current studies to date, based solely on hourly aggregates, with no justification. The proposed approach considered the optimal interval to be the one which maximizes the proportion of periods showing substantial statistical changes, and for this we used Pearson's

Chi-square and the Kolmogorov–Smirnov tests of goodness-of-fit.

We have established a procedure to determine the optimal intervals of traffic from traffic analysis. To determine the optimal period of analysis, we have studied the aggregate traffic from different periods (10, 15, 30, 45, 60, 90 and 120 min) using Pearson's Chi square and the Kolmogorov–Smirnov tests of goodness-of-fit, obtaining similar results from both tests.

The achieved results do not raise questions and we conclude that the optimal intervals are 15 min for uplink traffic, and 10 min for downlink traffic.

3

Comparative Analysis of Traffic Anomaly Detection Methods

In this chapter, we analyze how the different algorithms considered in this book (cumulative sum control chart, Pearson's Chi-squared and Kolmogorov–Smirnov (K-S) tests of goodness-of-fit, and mutual information (MI)) behave in detecting changing points.

In order to compare the traffic anomaly detection methods, we present a deep analysis of the monitored traffic to understand how the traffic behaves throughout the day.

3.1. Introduction

The change point detection theory is a subject that helps us to identify abrupt changes in a time series in general, and in network traffic in particular. As we present in section 3.2, according to the state of the art, the literature has not compared the different approaches of change point detection in depth, specifically the statistical control charts, the tests of goodness-of-fit or algorithms based on information entropy.

As an introduction to Chapter 4, the authors of the state of the art have focused on longitudinal traffic analysis (evolution in time), in order to identify sudden peak changes, rather than a 24 h typical day profile analysis, whose benefit lies in being able to determine traffic patterns within a daily profile.

The chapter is structured as follows: first, we provide a throughout state of the art analysis, and then we describe the methodology used. The chapter continues with the description of the proposed change point detection algorithms, and then the results and discussions, followed by the conclusions.

3.2. State of the art

One of the most important change point detection methods is the statistical control chart (SCC) algorithm, which was first introduced by W. A. Shewhart at Bell Labs in the 1920s under the concept of statistical process control (SPC) [SHE 24], and subsequent methods such as cumulative sum control chart (CUSUM) [BAS 93], developed at the University of Cambridge [PAG 54], and exponentially weighted moving average (EWMA) charts [COX 61].

In addition, there are other kinds of algorithms based on traffic statistical distributions which are also used to derive change point detection: the tests of goodness-of-fit.

In addition, we consider a third change point detection method called mutual information (MI) [SHA 49] as a statistical technique to measure the mutual dependence of traffic in different time intervals, which can also be used to detect changes in traffic.

The literature shows surveys that focus on different change point detection methods applied to either Internet Protocol (IP) networks in general [PEN 07] or on specific protocols or services, such as Voice over IP (VoIP) [EHL 10]. However, this is a theoretical study and the results are not evaluated empirically.

In [MAR 11] and [MAT 11], the authors made a comparison of SCC applied to IP traffic forecasts. The papers empirically describe the performance of EWMA, CUSUM and SPC for traffic trends, by using a longitudinal traffic analysis of 8 weeks to detect sudden peak changes. They conclude that EWMA obtains a better performance than CUSUM or SPC.

In addition, a later work [CAR 12] confirms that for different traffic series, SPC and CUSUM achieve the best performance, so the conclusion is that the performance of SCC depends on traffic behavior. Such conclusions are also supported by Oprea and Emile [OPR 13]. The comparisons obtained by the authors, again based on sudden peak changes, pinpoint that CUSUM and EWMA do not have the same suitability when it comes to practical implementations of their algorithms.

In [TAR 06], the authors show a comparison of SCC versus tests of goodness-of-fit. The work presents a benchmark of both methods when detecting intrusions in the network. In particular, the authors compare the CUSUM and EWMA algorithms with the Chi-squared test of goodness-of-fit. The experiments illustrate that the CUSUM algorithm detects the attack in a better way than the other algorithms. However, once again a longitudinal traffic analysis, namely, detecting sudden peak changes rather than typical day study, is made, and in addition the authors used detailed protocol-type information (e.g. Transmission Control Protocol (TCP), Synchronization bit (SYN), User Datagram Protocol (UDP) and Internet Control Message Protocol (ICMP) packets) with an accuracy of at least seconds, as most in the state of the art, that many times it is not available or accessible.

In [BUL 12], we can find the application of SCC for change point detection based on spectrum analysis rather than using raw traffic. The spectrum analysis is applied in some scenarios where the traffic is evaluated sequentially to detect sudden peak changes in order to get higher accuracy. The author concludes again that as with most data-driven approaches, there is no one change point detection technique that works on all types of data; different techniques perform better on different types of data. Yet, this work does not study typical day analysis, is based only on SCC and uses simulated time series data.

In the same research line [CAL 12], the authors propose to combine CUSUM with signal processing based on wavelets to detect sudden peak changes again. The experimental tests demonstrate the efficiency of the proposed solution for different traffic anomalies.

In [VER 08], the authors use the MI theory to improve the feature selection for fault detection. They present a fault diagnosis procedure based on discriminant analysis and MI. In order to obtain good classification performances, a selection of important features is done with a newly developed algorithm based on the MI between variables. Nevertheless, they do not use the MI theory as a criterion for fault detection. In the same way, in [AMI 11], the authors present MI-based feature selection for intrusion detection systems. The usefulness of MI for selecting the most relevant features in a given classification task has been proven in other fields [DRU 14].

In [SHA 06], the authors use the MI to implement a dynamic modeling of Internet traffic for intrusion detection. The results show that MI is especially useful in detecting flooding attacks such as constant bit rate (CBR) attacks.

Finally, in [MAT 14], the authors discuss the problem of anomaly detection in the situation that a strong trend is present in the traffic (diurnal pattern being a result of human behavior). As such a trend is a kind of non-stationarity by itself, its presence in most cases has a negative effect on the performance of any change point detection algorithm. To solve this, the authors propose a methodology for removing the inherent daily pattern. After removing the daily trend, they obtain standardized samples that are nearly normally distributed, as long as there is sufficient traffic aggregation as a consequence, the fit improves when the night periods are removed from the sample.

Thus, this paper exploits the presence of a weekly pattern to estimate and remove the seasonality from the measurements and proposes a methodology to simultaneously detect changes in mean and variance. The work relies on specific assumptions on the traffic type and nature as their study features VoIP calls, and therefore the methodology can only be applied to traffic following a non-homogeneous Poisson process. This means that they need to extract performance indices for each VoIP flow (voice calls) such as the call arrival process and call holding time distribution, and this information is not usually available. All in all, their work cannot detect changes in low traffic periods such as night-time.

This way, the state of the art in comparing the performance of change point detection algorithms is basically limited to SCC algorithms [MAR 11, MAT 11, CAR 12, OPR 13] and a few works broaden the benchmark to tests of goodness-of-fit (Pearson's Chi-squared test in [TAR 06]) but from protocol-type detailed traffic and to detect sudden peak changes. Some research lines propose change point detection based on spectrum analysis [BUL 12] or signal processing based on wavelets [CAL 12], but all the approaches are useful when the traffic is analyzed in terms of sudden peak changes rather than typical day profile (24 h aggregated traffic).

In addition, in [BAD 14], the authors review and evaluate the state-of-the-art studies on the problem of anomaly detection in computer networks. They provide an elaborate description of the anomaly detection problem, and depict the different categorizations of its solutions. They identified two levels for handling this problem; the network level and the application level. The network-level detection, on the one hand, analyzes the headers and/or the payloads of the messages exchanged in the network, while application-level detection analyzes the application specification and/or examines its behavior during runtime.

3.3. Average-day preliminary analysis

In our study, we have used a series of Internet traffic flows from a campus university with around 15,000 Internet users. We have analyzed 8 weeks of highly multiplexed traffic monitored downlink (from the Internet to the user) and uplink (from the user to the Internet) that was collected between October and December 2011.

We have focused on the link direction with higher traffic load (downloads) and in order to center the study of this work on the business days, we have not considered the traffic of the weekends.

Therefore, in our study, we use a total of 39 working days, downlink traffic (from the Internet to the user) monitored at the interface among different schools, added at intervals of 1 h to obtain a mean throughput (kB/s).

3.3.1. *Baseline scenario*

Figure 3.1 represents the monitored traffic of the working days (downlink) added hourly, in kB/s. This traffic forms the baseline scenario for evaluating the change point detection algorithms.

Next, we analyze the characteristics of the multiplexed traffic, and therefore the whole series of days for the same period have been analyzed. We have studied a cross-sectional analysis of the days considered in our study. Figure 3.2 represents the 24 h traffic shape of the different business days.

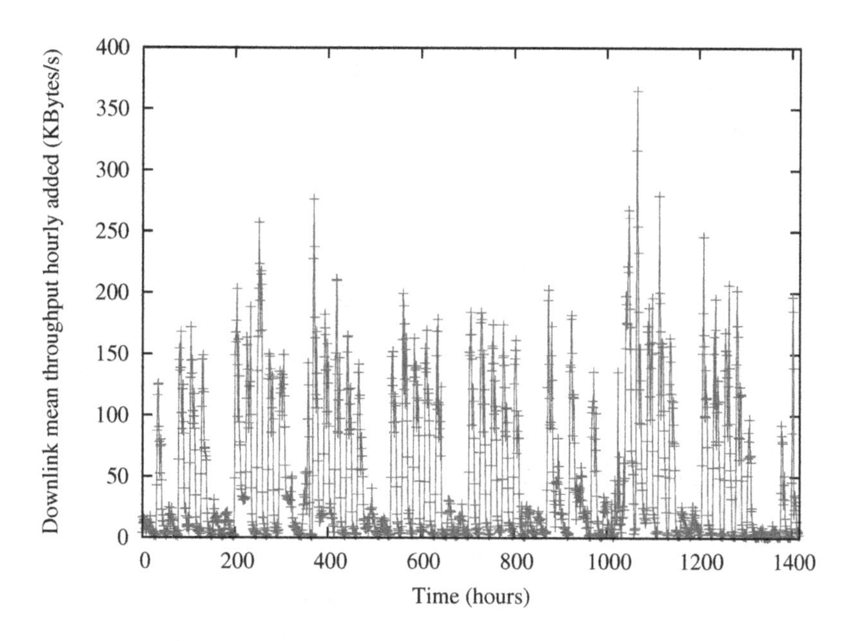

Figure 3.1. *Traffic profile of the monitored days (downlink, kB/s)*

3.3.2. *Anomalous scenario*

To test the behavior of the algorithms, we create an anomalous scenario on which we apply the algorithms, and then we compare the results obtained from the baseline scenario (i.e. the monitored traffic).

In order to create the anomalous scenario, we add a new day with abnormal traffic. The traffic shape by default of the 40th day is the average of the traffic values of each period.

Thus, the new day is contaminated in three well-differentiated periods:

1) low traffic (4am);

2) busy hour (12am);

3) high average traffic (5pm).

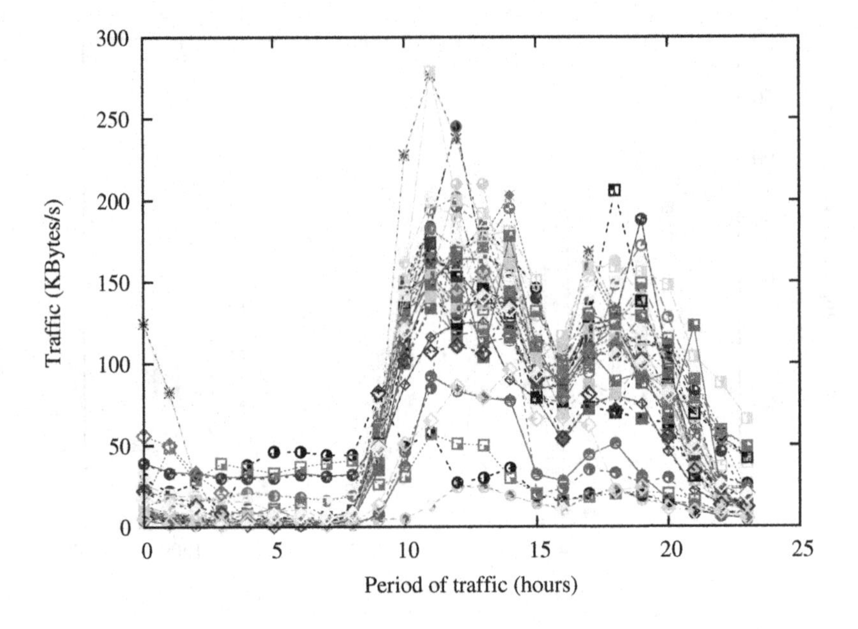

Figure 3.2. *Traffic profile of the 39 monitored working days. For a color version of the figure, see www.iste.co.uk/cuadra/traffic.zip*

Therefore, we inject traffic on those three intervals in terms of 40% higher than the maximum and 40% lower than the minimum of the entire series, and 40% higher than the maximum average and 40% lower than the minimum average of that period.

Thus, we compare the values of each algorithm before and after adding the contaminated traffic. Figure 3.3 shows the traffic contaminated evenhandedly and weighed on the 39 day baseline scenario (in gray).

3.4. Proposed change point detection algorithms

In this section, we gather the algorithms tested in this chapter: the most relevant statistical control chart (CUSUM), the two main tests of goodness-of-fit (Pearson's Chi-squared and Kolmogorov-Smirnov (K-S) tests) and MI. The study is based on the results included in [CUA 14] and an overview of the different algorithms can be found in Chapter 1.

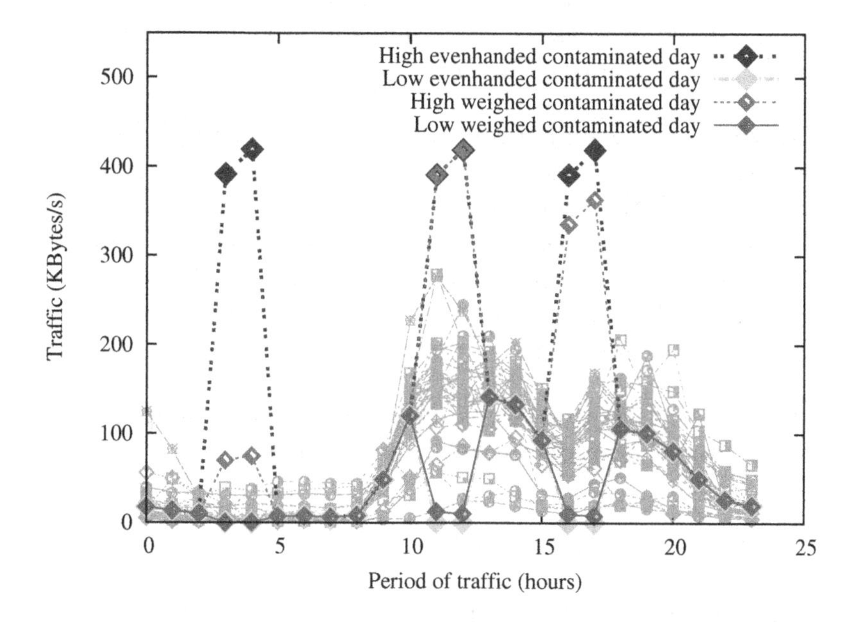

Figure 3.3. *Traffic contaminated evenhandedly and weighed. For a color version of the figure, see www.iste.co.uk/cuadra/traffic.zip*

3.4.1. *Statistical control charts*

The state of the art shows that there is no single SCC algorithm that stands out from the others, but the performance of SCC depends on the underlying traffic characteristics [MAR 11, MAT 11, CAR 12, OPR 13, BUL 12]. So, in order to simplify this study, we have focused on the most renowned one, the CUSUM chart, as the implementation of other control chart methods (e.g. EWMA) would lead to equivalent results.

Control charts perform measurements obtained as variations from the expected value by the standard deviation. In particular, CUSUM (CUMulative SUM) is a sequential analysis algorithm that allows us to monitor sudden changes in continuous processes, such as changes in traffic. The algorithm is summarized in Chapter 1.

The SCCs are graphs that show whether a sample of data falls within the normal range of variation. In CUSUM (also known as CSUM) as measurements are taken, the difference between each measurement and the benchmark value is calculated, and this is cumulatively summed up. When there is an anomaly in the traffic, the CUSUM value will progressively depart from that of the benchmark.

In this work, all the values from a period are compared to the ones from the immediate previous period.

3.4.2. *Tests of goodness-of-fit*

The tests of goodness-of-fit [CRO 12] are used to derive whether a given time interval of 1 day shows a traffic distribution independent from the rest of the intervals (i.e. how similar are two consecutive periods) that can be used to determine changes.

In this work, the two main tests of goodness-of-fit have been carried out. These tests are described in the following sections: Pearson's Chi-square test [CHE 54] and K-S test [PET 77].

3.4.2.1. *Pearson's Chi-squared test* (χ^2)

This test measures the discrepancies between the expected number of times each outcome occurs (assuming that the model is true) and the observed number of times each outcome occurs [PEA 00]. An overview of the algorithm is summarized in Chapter 1.

In this work, it is considered that the observed values correspond to the current aggregate time period, and expected the preceding period. To calculate the statistics of each interval, we create a histogram with the number of occurrences for different days.

These intervals have a similar probability distribution within these periods and independent of adjacent periods, well differentiated from the rest. This allows us to detect changes in network behavior in these time slots.

3.4.2.2. *Kolmogorov–Smirnov test*

This test [PET 77] checks that the distribution of a set of samples conforms to the theoretical distribution. The algorithm is summarized in Chapter 1.

To test the hypothesis, the two cumulative distribution functions (CDFs) are compared, that is the empirical and theoretical distribution function.

When comparing empirical values, the observed and expected CDFs are used, in this work the CDF of one period and the immediately preceding one. Once both distributions are calculated, the K-S statistic is determined from the biggest difference between both functions, the maximum distance between two consecutive CDFs.

3.4.3. *Mutual information*

The MI is a measure of the amount of information that one random variable contains about another one. This means that if both random variables are independent, the MI is zero. The algorithm is described in Chapter 1.

In this work, MI is used as a measure of the similarity between the traffic of two consecutive intervals. This way, the MI measures the amount of information that one period of traffic contains about the previous one and therefore constitutes an algorithm to detect changes in traffic.

Since there are no reference values in terms of figures that can take the MI algorithm, beyond the minimum is zero when they are independent variables, typically a value of normalized MI(MIn) with respect to the highest value that is taken in a series of traffic is used.

3.5. Behavior of the analyzed algorithms

In this section, we show the behavior of the different change point detection methods by comparing the results obtained on the baseline versus the anomalous scenario.

3.5.1. *Methodology*

We have tested the four algorithms (CUSUM, χ^2, K-S and MI) on the original traffic, which constitutes the baseline scenario, and then on the anomalous scenario with contaminated traffic, in which 1 new day is added to the original traffic in order to emulate that a new period of a day with changes has arrived.

In this section, we show the differences between the values taken by the algorithms with the original traffic and the contaminated traffic, measured as a percentage.

As we have introduced in the previous section, the traffic is contaminated in three specific periods (low traffic -4am, busy hour -12am and average high traffic -5pm) in terms of 40% higher than the maximum and 40% lower than the minimum of the entire series (evenhandedly), and 40% higher than the maximum average and 40% lower than the minimum average of that period (weighed).

Figures 3.4–3.7 show the results for different algorithms: CUSUM, χ^2, K-S and MI. The values of each algorithm represent the differences between the values obtained from the original traffic with respect to the contaminated traffic (whose shape is represented in gray).

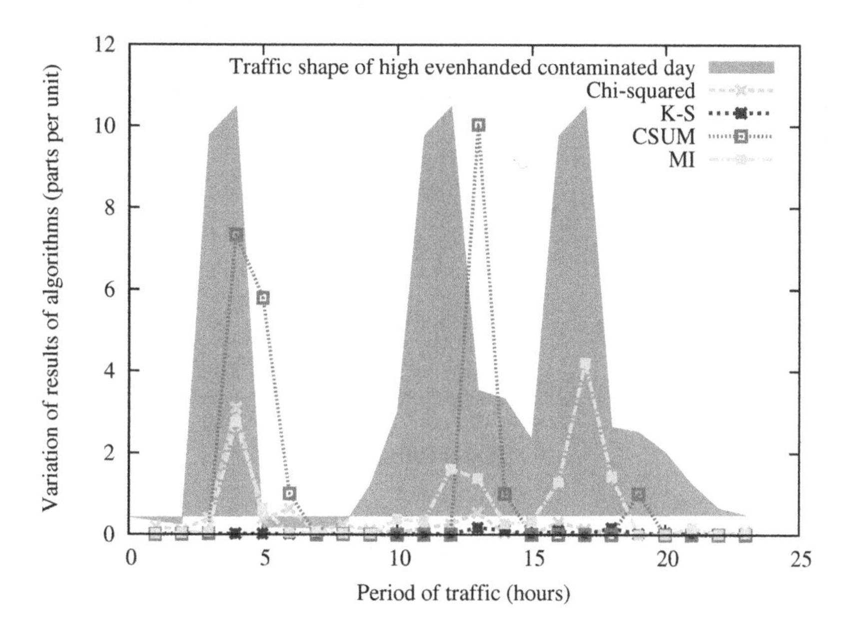

Figure 3.4. *High traffic contaminated (evenhanded). For a color version of the figure, see www.iste.co.uk/cuadra/traffic.zip*

To sum up, we added new fictitious days using as a baseline the mean value of the whole traffic series at each hour (see gray area on each figure), on which we have increased (or decreased) the traffic in those three periods to test the behavior in different traffic scenarios: low traffic (4am, nighttime), busy hour (12am, midday) and average high traffic (5pm, afternoon).

Also note that the ordinate axis represents the difference between the results of the algorithm obtained from the original traffic versus the same set adding the new day with contaminated traffic.

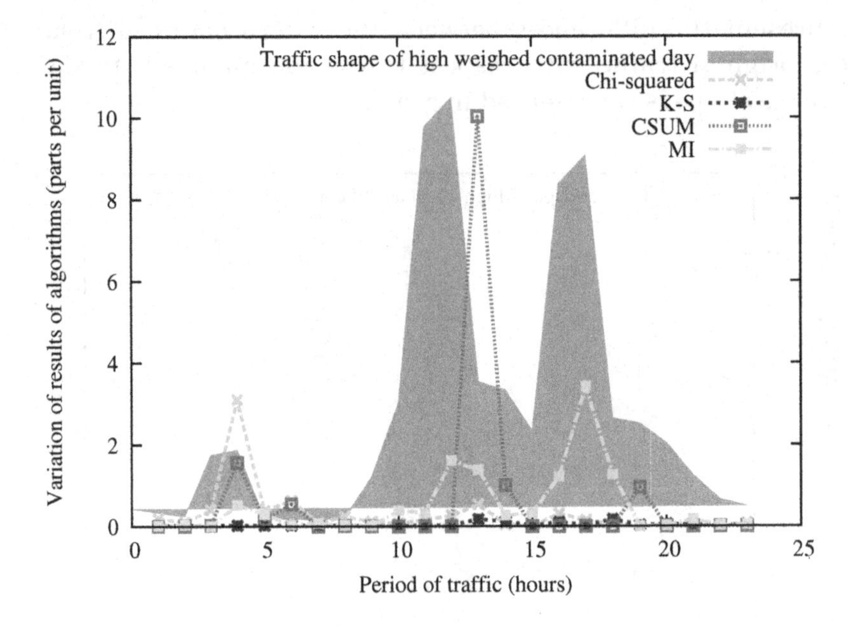

Figure 3.5. *High traffic contaminated (weighed). For a color version of the figure, see www.iste.co.uk/cuadra/traffic.zip*

First, we observe that the K-S algorithm is not able to detect any contaminated period but slightly detects the first contaminated period (4am) in the last two scenarios (Figures 3.4 and 3.5), so we have not taken it into account in our analysis and conclusions.

In the first scenario (Figure 3.4), we have significantly increased the traffic in the three periods with the same value at 4am, 12am and 5pm: 40% higher than the maximum traffic value of the entire series (425 KBytes/s). MI detects the three periods (an increase of 30, 20 and 40%, respectively, vs. the traffic not contaminated), CUSUM clearly detects the first two periods (70 and 100%) and χ^2 clearly detects the first period (an increase of 30%).

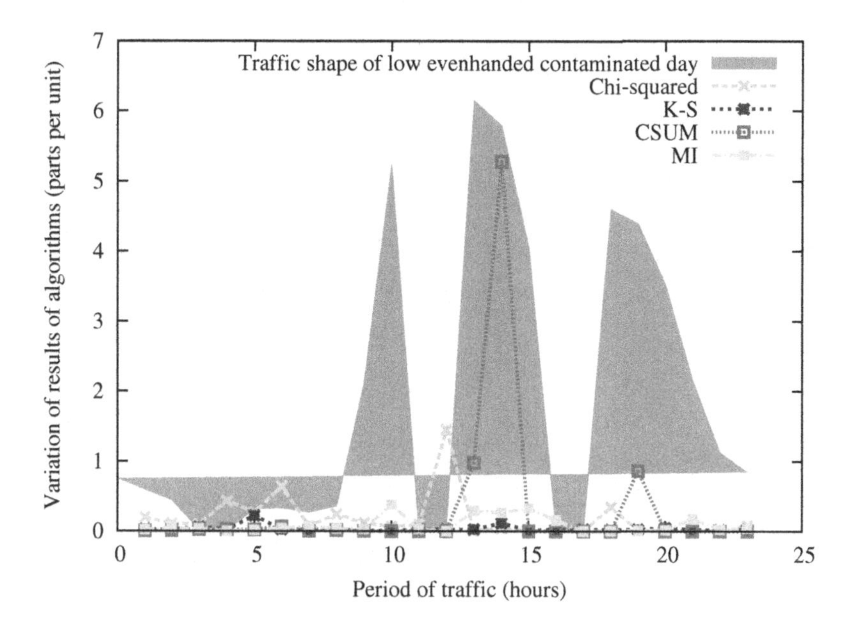

Figure 3.6. *Negligible traffic contaminated (evenhanded). For a color version of the figure, see www.iste.co.uk/cuadra/traffic.zip*

The second scenario (Figure 3.5) differs from the first scenario in that the increased traffic in those periods is not the same but 40% higher than the maximum average inside each period (75 KBytes/s at 4am, 425 KBytes/s at 12am and 350 KBytes/s at 5pm). MI detects the last two periods (20 and 40%), CUSUM detects the intermediate period (100% increase) and χ^2 detects the first period (30% higher).

In the third scenario (Figure 3.6), we have notably decreased the traffic 40% lower than the minimum of the entire series for the three periods evenhandedly (1 KByte/s). χ^2 detects the three periods (increase of 5, 15 and 4%, respectively), CUSUM clearly detects the last two periods (50 and 10%), meanwhile MI cannot significantly detect any contaminated period with contaminated low traffic.

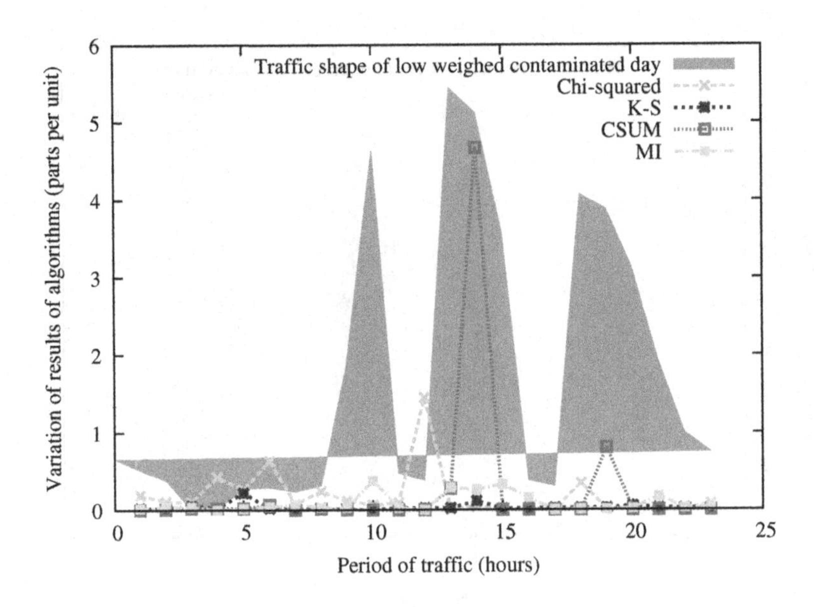

Figure 3.7. *Negligible traffic contaminated (weighed). For a color version of the figure, see www.iste.co.uk/cuadra/traffic.zip*

Finally, in the fourth scenario (Figure 3.7), we have decreased the traffic in the periods of the contaminated day 40% lower than the minimum average of each period (1 KByte/s at 4am, 10 KBytes/s at 12am and 8 KBytes/s at 5pm). The results are similar to the previous scenario as χ^2 detects the three periods contaminated (5, 15 and 4%), CUSUM detects the last two periods (50 and 10%, respectively) and MI cannot detect any.

From these results, we can summarize that:

– high contamination (maximum constant value):

- MIn detects the three variations (increasing 30, 20 and 40%, respectively),

- CSUM detects the three variations (increasing 70, 100 and 10%, respectively),

- χ^2 detects the first variation (increasing 30%);

– high contamination (maximum variable value):

- MIn detects the three variations (increasing 5, 20 and 40%, respectively),

- CSUM detects the variation in the middle (increasing 100%),

- χ^2 detects the first variation (increasing 30%);

– negligible contamination (minimum constant value):

- χ^2 detects the three variations (increasing 5, 15 and 4%, respectively),

- CSUM detects the last two variations (increasing 50 and 10%, respectively);

– negligible contamination (minimum variable value):

- χ^2 detects the three variations (increasing 5, 15 and 4%, respectively),

- CSUM detects the last two variations (increasing 50 and 10%, respectively).

The results of this comparison are analyzed in Chapter 4 to consolidate a new information-theory-based technique.

3.6. Conclusion

In this chapter, we have analyzed the implementation of the main change point detection methods in order to determine which one obtains the best results throughout the day.

This approach is more useful than sudden peak changes based on traditional longitudinal traffic analysis, as we consider the concrete traffic period in which the anomaly occurs, as traffic behaves differently throughout the day.

We have implemented four algorithms belonging to these methods, and we have tested them under real traffic conditions and under contaminated traffic to determine which one best detects those changes.

4

Proposal of a New Information-theory Technique

In this chapter, we propose a new information-theory technique based on the results analyzed in Chapter 3. The change-point detection philosophy helps us to detect changes produced in the network. The current techniques, according to the state of the art, are based on longitudinal traffic analysis to detect abrupt peak changes regardless of the time of the day.

Nevertheless, the traffic varies throughout the day and may experience different behaviors depending on the time of the day. We believe that considering the concrete traffic period in which the anomaly occurs will help us to discern in a clearer way the traffic anomalies in multimedia services. This new information-theory technique is what we call typical day analysis, and is based on the study of the traffic pattern over a typical 24 h day.

In this chapter, we analyze how the different traffic anomaly detection methods perform in detecting changing points inside a typical day profile. We conclude that there is not a single method that can be highlighted, but a combination of different algorithms depending on the underlying traffic and the time of the day which provides much more accuracy in detecting traffic changes.

4.1. Introduction

The state of the art regarding network management, and in particular related to change-point detection (see Chapter 3), is focused on longitudinal traffic analysis (evolution in time), in order to identify sudden peak changes, rather than a 24 h typical day profile analysis, which benefit lies in being able to determine traffic patterns within a daily profile.

The 24 h typical day analysis has obvious advantages over just detecting sudden peak changes. For example, a traffic peak may always occur at a certain time of the day and it may not be anomalous at all. It can only be considered anomalous if it is not "typical at that time of the day. Actually, the traffic consumption varies throughout the day, establishing a typical day profile with low traffic at nighttime, and high traffic during working hours, with a busy hour around midday.

In Chapter 3, we have compared the existing change-point detection techniques with an algorithm based on mutual information to identify abnormalities in subsequent time intervals. In this chapter, we determine the performance of the different methods when detecting traffic changes in a typical day and conclude that the accuracy of the change-point detection algorithms depends on traffic conditions.

4.2. Related work

The state of the art detailed in Chapter 3 shows that there are no studies that compare the main change-point detection methods in any depth: SCC, tests of goodness-of-fit and algorithms based on mutual information analysis.

Furthermore in the state of the art, there are no surveys that test change-point detection algorithms for a typical average day since all of them address the studies to detect sudden peak changes. Moreover, previous studies utilize very detailed traffic information, such as protocol type, and with an accuracy of seconds.

However, most of the monitoring data sources do not have such a level of detail of the protocol or even time precision in seconds, but are based on aggregated traffic information, typically in the order of minutes, where only the traffic rate is available.

We can conclude that no author to date has proposed to detect changing points inside a 24 h period traffic pattern, a new disruptive information-theory based technique that we call "typical day profile" which we present in the following sections. Moreover, we analyze how the different algorithms behave in detecting changing points inside a typical day profile.

4.3. Analysis of traffic anomaly detection methods applied to typical day profile

In Chapter 3, we made a comparative analysis of traffic anomaly detection methods. We tested our procedure on the data set from six weeks of highly multiplexed traffic monitored at a campus university.

In order to study the procedure under different anomalous traffic, we have added a new day constituting the network anomaly, in which we have inserted traffic in the following way. The traffic is contaminated in three specific periods (low traffic -4am-, busy hour -12am- and average high traffic -5pm-) in terms of 40% higher than the maximum and 40% lower than the minimum of the entire series (evenhandedly), and 40% higher than the maximum average and 40% lower than the minimum average of that period (weighed).

Therefore, to benchmark the values of the algorithms we added anomalous traffic (contaminated) in the following way:

– 40% higher than the maximum of the entire series, and 40% lower than the minimum of the entire series:

- high contamination (the same maximum value for the whole day),

- negligible contamination (the same minimum value for the whole day);

– 40% higher than the maximum average of that period, and 40% lower than the average minimum of that period:

- high contamination (different maximum value depending on the period of the day),

- negligible contamination (different minimum value depending on the period of the day).

The results applied to the typical day profile are summarized in Table 5.1, where the number in brackets means the variation (percentage) of the algorithms with contaminated traffic from the original traffic. The algorithms did not produce false positives, namely a false alarm.

Contaminate	*12am–1pm*	*5pm–6pm*	*4am–5am*
High Evenhanded	CSUM(100%) MI(20%)	MI(40%)	CSUM(70%) χ^2(30%) MI(30%)
High Weighed	CSUM(100%) MI(20%)	MI(40%)	χ^2(30%)
Low Evenhanded	CSUM(50%) χ^2(15%)	CSUM(10%) χ^2(4%)	χ^2(5%)
Low Weighed	CSUM(50%) χ^2(15%)	CSUM(10%) χ^2(4%)	χ^2(5%)

Table 4.1. *Summary of results for high and low contamination*

4.3.1. *Choice of algorithms*

Table 5.2 shows the best algorithm or combination of algorithms based on traffic rate. We can generalize that:

– MI: best to detect traffic rises on daytime traffic load intervals;

– χ^2: best to detect traffic drops at anytime (regardless the traffic rate), and traffic rises on nighttime traffic load intervals;

– CUSUM: best to detect very abrupt changes at anytime (regardless the traffic rate).

Traffic	*Normal (10am–10pm)*	*Low (11pm–9am)*
High contamination	MI (+CSUM)	χ^2 (+CSUM)
Low contamination	χ^2 (+CSUM)	χ^2 (+CSUM)

Table 4.2. *Optimal algorithm(s) per traffic rate*

We can conclude that a traffic change point detection solution should use the three previous algorithms in order to detect traffic changes properly: CUSUM for abrupt changes (1) and χ^2 to detect traffic drops (2) regardless of the traffic load (the whole day); MI to detect traffic rises under normal traffic conditions (3) while χ^2 in low traffic conditions (4).

– Regardless of the traffic load (24 h)

$$\text{Variation of CUSUM} > 50\% \rightarrow \text{Abrupt change} \quad [4.1]$$

$$\text{Variation of } \chi^2 > 5\% \rightarrow \text{Traffic drop} \quad [4.2]$$

– Normal traffic (10am–10pm)

$$\text{Variation of MI} > 20\% \rightarrow \text{Traffic rise} \quad [4.3]$$

– Low traffic (11pm–9am)

$$\text{Variation of } \chi^2 > 30\% \rightarrow \text{Traffic rise} \qquad [4.4]$$

The traffic anomaly detection analysis carried out in this work points out the combination of algorithms which gets a better performance for each traffic period of a typical day. For simplification, we have categorized the typical day into three main different periods (nighttime, daytime and busy hour) although this method can also be applied to other time periods, such as hours.

4.4. Conclusions

In this chapter, we have presented the "typical day profile" technique and its application to change point detection field. This constitutes a new information-theory based technique which analyzes the traffic pattern over a typical 24 h day.

After carrying out a deep traffic analysis to be aware of normal traffic behavior, we adjust our procedure by considering which combination of algorithms gets a better performance in each period of the day.

Moreover, we have evaluated the procedure using six weeks of traffic from a campus university, to which we have added a threat day by including contaminated traffic (flooding and absence of traffic) at three specific time periods: low traffic (4am, nighttime), busy hour (12am, midday) and average high traffic (5pm, afternoon).

The analysis we have performed confirms that a combination of the algorithms provides better results than the use of a single algorithm. In particular, for the network traffic considered in this work, three algorithms present a higher accuracy depending on the traffic and the type of anomaly introduced in the network.

Pearson's χ^2 test of goodness-of-fit should be used to detect anomalies related to traffic drops regardless of the time of the day, as well as anomalies linked to traffic increases at night. However, the Mutual Information algorithm best detects traffic rises during daytime. In addition, the CUSUM statistical control chart complements both of them when detecting very abrupt changes, regardless of the time of the day.

4.5. Acknowledgments

This work is carried out within the EUREKA Celtic-plus project Next generation Over-The-Top multimedia services (NOTTS) and co-funded by the Spanish public authority CDTI.

Bibliography

[AMI 11] AMIRI F., REZAEI YOUSEFI M., LUCAS C. *et al.*, "Mutual information-based feature selection for intrusion detection systems", *Journal of Network and Computer Applications*, Elsevier, vol. 34, no. 4, pp. 1184–1199, 2011.

[BAD 14] BADDAR S.A.-H., MERLO A., MIGLIARDI M., "Anomaly detection in computer networks: a state-of-the-art review", *Journal of Wireless Mobile Networks, Ubiquitous Computing, and Dependable Applications (JoWUA)*, vol. 5, no. 4, pp. 29–64, 2014.

[BAS 93] BASSEVILLE M., NIKIFOROV I.V. *Detection of Abrupt Changes: Theory and Application*, Prentice Hall, Englewood Cliffs, vol. 104, 1993.

[BEI 04] BEITZEL S.M., JENSEN E.C., CHOWDHURY A. *et al.*, "Hourly analysis of a very large topically categorized web query log", *Proceedings of the 27th Annual International ACM SIGIR Conference on Research and Development in Information Retrieval*, ACM, pp. 321–328, 2004.

[BEI 07] BEITZEL S.M., JENSEN E.C., CHOWDHURY A. *et al.*, "Temporal analysis of a very large topically categorized web query log", *Journal of the American Society for Information Science and Technology*, Wiley Online Library, vol. 58, no. 2, pp. 166–178, 2007.

[BEL 62] BELL C. *et al.*, "Mutual information and maximal correlation as measures of dependence", *The Annals of Mathematical Statistics*, Institute of Mathematical Statistics, vol. 33, no. 2, pp. 587–595, 1962.

[BRO 02] BROWNLEE N., CLAFFY K., "Understanding Internet traffic streams: dragonflies and tortoises", *Communications Magazine, IEEE*, vol. 40, no. 10, pp. 110–117, 2002.

[BUL 12] BULUNGA M.L., Change-point detection in dynamical systems using auto-associative neural networks, PhD Thesis, Stellenbosch University, 2012.

[CAL 10] CALLAHAN T., ALLMAN M., PAXSON V., "A longitudinal view of HTTP traffic", *Passive and Active Measurement*, Springer, pp. 222–231, 2010.

[CAL 12] CALLEGARI C., GIORDANO S., PAGANO M. *et al.*, *WAVE-CUSUM*, Computers Security, 2012.

[CAR 12] CARVALHO A.M.M., Controle estatéstico de processos de predição de tráfego de redes de computadores, Master's Thesis, University of Uberlândia, 2012.

[CHE 54] CHERNOFF H., LEHMANN E., "The use of maximum likelihood estimates in χ^2 tests for goodness of fit", *The Annals of Mathematical Statistics*, Institute of Mathematical Statistics, vol. 25, no. 3, pp. 579–586, 1954.

[COX 61] COX D.R., "Prediction by exponentially weighted moving averages and related methods", *Journal of the Royal Statistical Society*, vol. series B (methodological), pp. 414–422, 1961.

[COX 66] COX D.R., LEWIS P.A., "The statistical analysis of series of events", *Monographs on Applied Probability and Statistics*, Chapman and Hall, London, vol. 1, 1966.

[CRO 10] CROARKIN C., TOBIAS P., NATRELLA M. *et al.*, NIST/SEMATECH e-Handbook of Statistical Methods, available at: http://www.itl.nist.gov/div898/ handbook, July 2010.

[CRO 12] CROARKIN C., GUTHRIE W., NIST/SEMATECH e-Handbook of Statistical Methods, National Institute of Standards and Technology (NIST), 2012.

[CUA 14] CUADRA-SANCHEZ A., ARACIL J., DE SANTIAGO J.R., "Proposal of a new information-theory based technique and analysis of traffic anomaly detection", *International Conference on, Smart Communications in Network Technologies (SaCoNeT)*, IEEE, pp. 1–6, 2014.

[DRU 14] DRUGMAN T., "Using mutual information in supervised temporal event detection: application to cough detection", *Biomedical Signal Processing and Control*, Elsevier, vol. 10, pp. 50–57, 2014.

[EHL 10] EHLERT S., GENEIATAKIS D., MAGEDANZ T., "Survey of network security systems to counter SIP-based denial-of-service attacks", *Computers & Security*, vol. 29, no. 2, p. 225–243, 2010.

[FOM 04] FOMENKOV M., KEYS K., MOORE D. *et al.*, "Longitudinal study of Internet traffic in 1998-2003", *Proceedings of the Winter International Symposium on Information and Communication Technologies*, Trinity College, Dublin, pp. 1–6, 2004.

[GAR 11] GARCÍA-DORADO J.L., HERNÁNDEZ J.A., ARACIL J. *et al.*, "Characterization of the busy-hour traffic of IP networks based on their intrinsic features", *Computer Networks*, Elsevier, vol. 55, no. 9, pp. 2111–2125, 2011.

[HIG 04] HIGGINS J.J., *An Introduction to Modern Nonparametric Statistics*, Brooks/Cole, Pacific Grove, CA, 2004.

[MAD 06] MADHUKAR A., WILLIAMSON C., "A longitudinal study of P2P traffic classification", *14th IEEE International Symposium on Modeling, Analysis, and Simulation of Computer and Telecommunication Systems, MASCOTS*, IEEE, pp. 179–188, 2006.

[MAI 09] MAIER G., FELDMANN A., PAXSON V. *et al.*, "On dominant characteristics of residential broadband internet traffic", *Proceedings of the 9th ACM SIGCOMM Conference on Internet Measurement Conference*, ACM, pp. 90–102, 2009.

[MAR 11] MARIA A., MATIAS R., MACEDO A. *et al.*, "Performance analysis of control charts techniques applied to IP traffic forecasts", *IEEE 12th International Conference on Parallel and Distributed Computing, Applications and Technologies (PDCAT)*, pp. 109–115, 2011.

[MAR 12] MARINESCU D.C., MARINESCU G.M., *Classical and Quantum Information*, Academic Press, 2012.

[MAT 11] MATIAS R., CARVALHO A.M., ARAUJO L.B. *et al.*, "Comparison analysis of statistical control charts for quality monitoring of network traffic forecasts", *IEEE International Conference on Systems, Man, and Cybernetics (SMC)*, pp. 404–409, 2011.

[MAT 14] MATA F., URANIEWSKI P., MANDJES M. *et al.*, "Anomaly detection in diurnal data", *Computer Networks*, Elsevier, vol. 60, pp. 187–200, 2014.

[MCC 00] MCCREARY S., CLAFFY K., Trends in wide area IP traffic patterns. A view from ames Internet exchange, CAIDA Report, 2000.

[MON 04] MONTGOMERY D., *Introduction to Statistical Quality Control*, Wiley, 2004.

[OPR 13] OPREA R., EMILE A., Traffic Anomaly Detection Using a Distributed Measurement Network, University of Amsterdam, 2013.

[PAG 54] PAGE E.S., "Continuous inspection schemes", *Biometrika*, vol. 41, no. 1/2, pp. 100–115, June 1954.

[PAP 05] PAPAGIANNAKI K., TAFT N., ZHANG Z.-L. *et al.*, "Long-term forecasting of Internet backbone traffic", *IEEE Transactions on Neural Networks,* IEEE, vol. 16, no. 5, pp. 1110–1124, 2005.

[PAX 96] PAXSON V., "End-to-end routing behavior in the Internet", *ACM SIGCOMM Computer Communication Review*, ACM, vol. 26, pp. 25–38, 1996.

[PEA 00] PEARSON K., "On the criterion that a given system of deviations from the probable in the case of a correlated system of variables is such that it can be reasonably supposed to have arisen from random sampling", *The London, Edinburgh, and Dublin Philosophical Magazine and Journal of Science*, Taylor & Francis, vol. 50, no. 302, pp. 157–175, 1900.

[PEN 07] PENG T., LECKIE C., RAMAMOHANARAO K., "Survey of network-based defense mechanisms countering the DoS and DDoS problems", *ACM Computing Surveys*, vol. 39, no. 1, p. 3, 2007.

[PET 77] PETTITT A.N., STEPHENS M.A., "The Kolmogorov-Smirnov goodness-of-fit statistic with discrete and grouped data", *Technometrics*, Taylor & Francis, vol. 19, no. 2, pp. 205–210, 1977.

[SHA 49] SHANNON C., WEAVER W., *The Mathematical Theory of Communication*, Illinois University Press, 1949.

[SHA 01] SHANNON C.E., "A mathematical theory of communication", *ACM SIGMOBILE Mobile Computing and Communications Review*, ACM, vol. 5, no. 1, pp. 3–55, 2001.

[SHA 06] SHAH K., JONCKHEERE E., BOHACEK S., "Dynamic modeling of internet traffic for intrusion detection", *EURASIP Journal on Advances in Signal Processing*, Hindawi Publishing Corporation, vol. 2007, no. 1, p. 090312, 2006.

[SHE 24] SHEWHART W.A., "Some applications of statistical methods to the analysis of physical and engineering data", *Bell System Technical Journal*, vol. 3, no. 1, pp. 43–87, 1924.

[TAR 06] TARTAKOVSKY A.G., ROZOVSKII B.L., BLAZEK R.B. *et al.*, "A novel approach to detection of intrusions in computer networks via adaptive sequential and batch-sequential change-point detection methods", *IEEE Transactions on Signal Processing*, vol. 54, no. 9. pp. 3372–3382, 2006.

[VER 08] VERRON S., TIPLICA T., KOBI A., "Fault detection and identification with a new feature selection based on mutual information", *Journal of Process Control*, Elsevier, vol. 18, no. 5, pp. 479–490, 2008.

[WIL 98] WILLINGER W., PAXSON V., TAQQU M.S., "Self-similarity and heavy tails: structural modeling of network traffic", *A Practical Guide to Heavy Tails: Statistical Techniques and Applications*, Birkhauser, Boston, MA, vol. 23, pp. 27–53, 1998.

Index

Printed in the United States
By Bookmasters